TRANSACTIONS

of the

American Philosophical Society

Held at Philadelphia for Promoting Useful Knowledge

VOLUME 77, Part 7, 1987

Al-Kāshī's Geographical Table

E. S. KENNEDY

and

M.-H. KENNEDY

Institut für Geschichte der Arabisch Islamischen
Wissenschaften, Frankfurt a. M.

THE AMERICAN PHILOSOPHICAL SOCIETY

Independence Square, Philadelphia

1987

Library of Congress Catalog
Card Number 87-72865
International Standard Book Number 0-87169-778-5
US ISSN 0065-9746

CONTENTS

INTRODUCTION

Jamshīd Ghyāth al-Dīn al-Kāshī (fl.1420), a native of central Iran, worked at the Samarqand observatory of Sultan Ulugh Beg, grandson of Tamerlane. Kāshī is best known as a virtuoso of computational mathematics, but he was also a competent astronomer. Among his major works is the *Zīj-i Khāqānī*, an astronomical handbook written in Persian. The zīj contains a list giving the latitudes and longitudes of some 515 places, mostly cities, the subject of this paper.

Such lists were common in the world of medieval Islam. The present author has collected and stored on magnetic tape the contents of some sixty-three sources comprising well over twelve thousand entries, and the number is steadily increasing. Most of these tables are found in unpublished manuscripts, but a few have been printed. Among them is the "Picture of the Earth" (*Ṣūrāt al-arḍ*, [*Khu.*]*) by al-Khwārizmī (fl. 825), which was the basis of the lost world map made for the caliph al-Maʾmūn. Another is found in the zīj of al-Battānī (fl. 900, [*Bat.*]), and a third in al-Bīrūnī's Masudic Canon (c.1035), [*Bir.*]). To our knowledge, however, no late medieval example has previously been published.

Reproduced in facsimile following page 40 is the India Office [*IO*] manuscript of the text, by kind permission of the Head of Photographic Administration, The British Library. In the original the cities are listed according to increasing "climates" (the concept is explained below), and within the climate ordered and numbered by increasing latitudes.

The list has also been transcribed into Latin characters, collated with the Aya Sofya copy, [*AS*], and arranged alphabetically just below. Where a name in the text differs considerably from its customary European form, the latter has also been introduced, as well as a cross reference from one to the other. Letters enclosed in square brackets are restorations, almost invariably cases in which a scribe has misplaced or omitted dots in the Arabic script.

The second column gives longitude followed by latitude, degrees being separated from minutes by a semicolon. In the text, the numerals are those of the Arabic alphabetical *abjad* system standard for scientific writings. Where a place can be located with reasonable probability, its modern

---

* Italicized short titles, sometimes enclosed in square brackets, are references to the bibliography at the end of the paper.

coordinates are shown enclosed in parentheses immediately below the medieval ones. This is the case for 382 cities, three-quarters of the total.

The third column enables the reader to locate any city in the original text. The first two digits, followed by an *r* or a *v*, identify the folio. After the colon a single digit followed by a comma gives the climate. The number after the comma is the place of the locality in the climate list.

The final column is used to locate the place, unless it is well known, to cite the authority justifying the location, or to make any speculations which seem reasonable.

THE TABLE

Place Name	Coordinates		Reference	Remarks
Ābā	85;10, (49;12,	34;40 35;35)	74r:4,103	Also Avah, mod. Avej, in Iran between Hamadan and Qazvin, *LS*, p. 196.
ʿAbādān	84;30, (48;15,	30;0 30;20)	73r:3,77	In Iran, on the Gulf.
Abarqūh	87;20, (53;18,	31;30 31;9)	73r:3,84	In central Iran, NNE of Shiraz.
Abaskūn, chief city of Jurjān.	89;30, (54;12,	37;15 37;30)	74r:4,116	Mod. Adzhiyap (*LS*, p. 376), in the Kazakhistan SSR, SE corner of the Caspian.
Abhar	84;30, (49;1,	36;45 36;5)	73v:4,97	In NW Iran, south of Rasht.
Abydos (text: Abzū)	59;45, (29;11,	44;0 40;25)	74v:6,8	Mod. Umurbey, Canakkale, near Istanbul.
Abzū, see Abydos				
Acre (text: ʿAkkā)	68;20, (35;4,	33;20 32;55)	73r:3,46	Mod. ʿAkko
Adana	69;15, (35;19,	36;50 37;0)	73v:4,31	In SE Turkey, near the Mediterranean.
Aden	76;0, (45;3,	11;0 12;50)	72v:0,14	Near the SW corner of the Arabian Peninsula.
Ahvāz	84;0, (48;43,	31;0 31;17)	73r:3,71	In the SW corner of Iran.
Akhlāṭ	75;50, (42;28,	39;20 38;45)	74r:5,34	Mod. Ahlat, in eastern Turkey on Lake Van.
Akhmīm	61;30, (31;48,	27;15 26;35)	72v:2,7	On the Nile, upstream from Asyut.
Akhsikat, capital of Farghāna	101;20,	42;25	74v:5,68	Near Namangan to the SW (*LS*, p. 477), in the Uzbekistan SSR. These coordinates are from the Aya Sofya MS. The India Office copy gives 101;35, 42;0.
ʿAkkā, see Acre				
Akla (?)	88;0,	49;0	74v:7,12	This place is named by no other source. It would seem to be somewhere west of the Caspian,

Place Name	Coordinates		Reference	Remarks
				east of Saratov (Saray). The longitude can also be read as 108°, which would put the location in Central Asia.
Alān, see Sarīr Alān				
ʿAlāya	62;0, (32;2,	39;30 36;32)	74r:5,22	Mod. Alanya (*LS*, pp. 142–150) in SE Turkey on the Mediterranean.
Aleppo (text: Ḥalab)	72;10, (37;10,	35;50 36;14)	73v:4,47	In the NW corner of Syria.
Alexandria	61;54, (29;55,	30;58 31,13)	73r:3,28	
Algeciras (text: [Kh]aḍrā' Island)	19;15, (−5;27,	35;50 36;8)	73v:4,6	In Spain, near Gibraltar.
ʿAllāqī	68;40,	27;15	72v:2,16	Other sources report this place as being near the Red Sea.
Amāsya	67;30, (35;50,	45;15 40;37)	74v:6,13	In north-central Turkey.
Āmid	77;20, (40;14,	37;52 37;55)	73v:4,68	Mod. Diyarbakir, in SE Turkey (*LS*, pp. 108–111).
ʿAmmān	67;20, (35;56,	31;0 31;57)	73r:3,42	The capital of Jordan.
ʿAmmūriya	64;0, (31;30,	43;0 39;15)	74r:5,23	Anc. Amorium, near mod. Assar-köy (*LS*, p. 135), east of Kütahya in western Turkey.
Amul, chief city of Ṭabaristān	87;20, (52;24,	36;0 36;26)	74r:4,111	Mod. Amol, in east-central Iran, near the Caspian.
ʿĀna	76;30, (41;57,	34;0 34;29)	73v:4,63	Mod. Anah, on the Euphrates NW of Baghdad.
Anbār	79;30, (43;46,	33;15 33;21)	73r:3,55	Mod. Fallujah (*LS*, p. 65), west of Baghdad.
Andarāb	103;45,	36;0	74r:4,150	According to *LS*, p. 427, in Afghanistan, east of Balkh.
Andījān, see Farghāna				
Ankūriya, called Anqara	64;40, (32;50,	41;0 39;55)	74r:5,24	Mod. Ankara, capital of Turkey.
Anṣuba, see Idfu				
Antioch (text: Anṭākīya)	71;26, (36;10,	35;40 36;12)	73v:4,43	Mod. Antakya, near the NE corner of the Mediterranean.
ʿAqaba of the sea of Egypt.	59;0,	32;0	73r:3,26	Probably not the Aqaba of Jordan.

Place Name	Coordinates		Reference	Remarks
Aqjākirmān	60;0, (30;19,	50;0 46;10)	74v:7,4	Mod. Belgorod Dnestrovskiy, in the Ukraine, near Odessa.
Aqrīṭish, see Crete				
Aqsaray	67;55, (34;2,	40;15 38;22)	74r:5,29	Mod. Aksary, Niğde, in central Turkey, SSE of Ankara.
Aqshahr	65;15, (31;24,	41;40 38;22)	74r:5,26	Mod. Akşehir, in west-central Turkey, SSW of Ankara.
Araq(?)	85;0,	48;0	74v:7,10	This locality would seem to be west of the Caspian, near mod. Saratov.
Arbūna, see Narbonne				
Ardabīl	82;20, (48;18,	37;20 38;15)	73v:4,85	In NW Iran, east of Tabrīz.
Arjīsh	77;0, (43;18,	38;30 39;0)	73v:4,66	Mod. Ercis in eastern Turkey, north of Lake Van.
Arrajān	84;30, (50;18,	32;30 30;37)	73r:3,73	In SW Iran, just north of Behbehān (*LS*, p. 268).
Arskluran(?)	106;0,	46;0	74v:6,25	This locality, given only in the late Persian zījes seems to be in Turkish Central Asia.
Arūdjard(?) Islands	——,	62;0	74v:8,4	Other sources give latitude 19;30° for the Abardaj(?) or Abūdū(?) Islands, with the same latitude given here. Probably the Hebrides. The mythical Thule has nearby coordinates.
Arzan-i Rūm	78;0, (41;17,	41;15 39;57)	74r:4,37	Mod. Erzurum in NE Turkey.
Arzinjān	78;0, (39;30,	39;50 39;44)	74r:5,36	Mod. Erzincan in east-central Turkey.
Asfī	12;15, (−9;20,	30;15 32;18)	73r:3,1	Mod. Safi, Morocco.
ʿAskar Mukram	84;30, (48;54,	31;15 31;40)	73r:3,74	Mod. Bandi-i Qīr (*LS*, p. 237), in SW Iran north of Ahvāz.
ʿAsqalān	66;30, (34;35,	32;0 31;40)	73r:3,39	Or Ascalon, or Ashqelon, in Palestine.
Astarābād	89;35, (54;29,	36;50 36;50)	74r:4,117	Mod. Gorgan in NW Iran.
Aswān	65;0, (32;56,	22;30 24;5)	72v:2,11	In southern Egypt, on the Nile.
Athens, City of Sages	65;40, (23;44,	37;20 38;0)	73v:4,26	
Aṭrāblus, see Tripoli				

Place Name	Coordinates		Reference	Remarks
Audghast	25;15,	26;0	72v:2,4	In the Maghrib, between Tlemçen and Ghana, according to *Abū'l-F.*, text, p. 136.
Avej, see Āba.				
Awāl Island	86;15, (50;32,	22;0 26;0)	72v:2,33	Mod. Uwal, on the Arabian side of the Gulf (cf. *LS*, p. 261), the main island of Bahrayn.
Awjān	81;30, (47;5,	37;20 37;29)	73v:4,80	Mod. Hashtrūd (*LS*, p. 163), in NW Iran, SSE of Tabriz.
Ayās	69;17, (35;45,	36;15 36;46)	73v:4,30	Mod. Yumurtalik, on the Mediterranean at its NE corner.
ʿAydhāb	68;0, (36;30,	21;0 22;20)	72v:2,15	On the African coast of the Red Sea, opposite Jiddah (*Maq.* pp. 80, 155).
Ayla	66;15, (34;57,	29;15 29;33)	73r:3,36	Mod. Elat, on the Gulf of Aqaba.
Aywān Kisrā, see Ctesiphon				
Baʿalbák	70;45, (36;12,	33;15 34;0)	73r:3,50	In the Lebanon.
Bāb al-Abwāb	85;0, (48;18,	46;0 42;3)	74r:5,45	Mod. Derbent, on the Caspian.
Bāb al-Ḥadīd	76;0, (67;0,	41;0 38;15)	74r:5,35	Mod. Derbent (*LS*, p. 441), in the Uzbekistan SSR south of Samarqand.
Bāb Iskandariya, see Iskenderun				
Babylon of ʿIraq	80;15, (44;25,	30;15 32;33)	73r:3,60	
Badakhshān	104;25, (70;40,	37;10 37;5)	74r:4,151	Mod. Faizabad (*LS*, p. 435), in NE Afghanistan.
Bādghīs	94;30,	35;20	74r:4,133	A region in Afghanistan NE of Herat (*LS*, pp. 412–415).
[B]adlīs	75;35, (42;4,	38;45 38;23)	73v:4,61	Mod. Bitlis, in Turkey SW of Lake Van (*LS*, p. 184).
Bāfd	90;15 (56;36,	29;15 29;12)	73r:3,92	Mod. Bāft, in SE Iran, south of Kerman.
Baghdād	80;15, (44;26,	33;20 33;20)	73r:3,58	
Baghrās	70;15, (36;11,	35;13 36;25)	73v:4,35	Near mod. Belen, in SE Turkey SE of Iskanderun, a medieval castle.
Baghshūr	96;25, (62;32,	36;0 35;40)	74r:4,137	Mod. Kalai-Mor (*LS*, p. 413), in the Turkmenia SSR, south of Mary (Merv).

Place Name	Coordinates		Reference	Remarks
Bahnasā	61;30, (30;40,	28;15 28;32)	73r:3,27	In Egypt, ancient Oxyrinchos, between Cairo and Asyut.
Baḥrayn, the last (part).	84;20, (50;38,	25;12 26;12)	72v:2,32	
Bāja	39;45, (9;13,	31;15 36;43)	73r:3,14	Mod. Beja, Tunisia.
Bajja of Berber	65;0,	14;0	72v:1,6	Presumably in NE Africa, possibly the place named in *Ḥud.*, p. 164.
Bajmāna(?)	32;15,	24;15	72v:2,5	In the Maghrib. The text reading is a conjecture. No other source gives this place.
Bakrābād, see Takīnābad.				
Bākūba (Baku)	84;30, (49;53,	39;30 40;22)	74r:5,42	On the western coast of the Caspian.
Balad	76;40, (42;43,	37;40 36;31)	73v:4,64	Mod. Eski Mosul (*LS*, p. 99) in northern Iraq NW of Mosul.
Balanjar, Khazar capital	85;0,	46;30	74v:6,17	A town somewhere in the Caucasus. (*Ḥud.*, p. 452).
Bālāsāghūn	101;30,	37;40	74r:4,144	Sometime capital of the Turks, the site unknown, near Kashgar beyond the Sirdarya (Jaxartes, *LS*, p. 487).
Balkh, the Cupola of Islam	101;0, (66;50,	36;41 36;46)	74r:4,142	In north-central Afghanistan, west of Mazār-i Sharīf.
Bam	94;0, (58;20,	28;30 29;7)	73r:3,99	In SE Iran, SE of Kerman.
Bāmiyān	102;0, (67;45,	34;35 34;52)	74r:4,145	In east-central Afghanistan, NW of Kabul.
Banī Kāwān Island	88;0, (56;17,	25;0 26;58)	72v:2,34	Mod. Qeshm, in the Gulf, cf. *LS*, p. 261.
Banjhīr (or Panjshīr) of Kābulistān	104;40, (69;51,	35;0 35;40)	74r:4,153	In east-central Afghanistan, north of Kabul.
Banjū(?)	135;0,	22;0	73r:2,47	Text: seat of the king of China. Many sources give essentially the same coordinates.
Barcelona (text: Barshalūna)	34;30, (2;10,	42;0 41;25)	74r:5,13	On the Mediterranean coast of Spain.
Barda'a	83;0, (47;8,	40;30 40;23)	74r:5,38	Mod. Barda (*LS*, p. 177), in the Azerbaydzhan SSR, near the Kur River west of Baku.

Place Name	Coordinates		Reference	Remarks
Bardān	79;50,	33;30	73r:3,57	According to *Abū'l-F.*, (transl., vol. 2, pt. 2, p. 75) near Baghdad.
Bardsīr	92;30,	30;15	73r:3,97	*LS*, p. 300, states that this place is the modern Kirmān. But Kāshī gives the latter, with coordinates differing from those of Bardsīr.
Barīsā(?)	32;0,	20;35	72v:1,2	In Africa, north of the Niger according to *Abū'l-F.* (transl., vol. 2, pt. 1, p. 220) in the Takrūr country.
Barqa	52;0, (21;54,	30;15 32;48)	73r:3,23	Mod. Shaḥḥat, Libya.
Barsbart	69;20 (40;16,	37;55 40;15)	73v:4,33	Mod. Bayburt, in NE Turkey, NW of Erzerum.
Baṣra	84;0, (47;50,	30;0 30;30)	73r:3,69	In Iraq, at the head of the Gulf.
Baykand	97;30, (64;15,	39;45 39;46)	74r:5,16	Near mod. Sverdlovsk (*LS*, p. 463), in the Uzbekistan SSR, a suburb of Bukhārā.
Baylaqān	83;30, (47;6,	39;50 39;46)	74r:5,41	A city in Azarbaijan which no longer exists. The site is near modern Martini, midway between Baku and Yerevan.
Beirut	69;30, (35;30,	34;0 33;52)	73v:4,29	
Beisān	68;15, (35;30,	32;50 32;30)	73r:3,45	Or Beyt Shean.
Benares	117;20, (83;0,	26;15 25;20)	73r:2,46	Mod. Varanasi.
Berbara of Zanj	78;0, (45;2,	6;30 10;28)	72v:0,15	In Somalia, on the coast.
Bilāsaḥriya(?)	65;0,	6;0	72v:0,8	Presumably somewhere in central Africa; given by no other source.
Bimānshahr	84;13,	37;30	73v:4,95	A village, site unknown, in the Iranian province of Daylam (*LS*, p. 174).
Binkath, capital of Shāsh	101;0, (69;13,	43;0 41;16)	74v:5,66	Mod. Tashkent (*LS*, p. 482) in the Uzbekistan SSR.
Bīrūn, see Nīrūn				
Bīshbāliq	111;0, (89;10,	45;50 44;1)	74v:6,28	Near mod. Jimsar, in the Xinjiang (Sinkiang) Uygur Zizhiqu, China, east of Urūmqi (*Ḥud.*, p. 272).

Place Name	Coordinates		Reference	Remarks
Biskra	32;40, (5;41,	30;15 34;50)	73r:3,9	In Algeria, SE of Algiers.
Bisṭām	89;30, (55;3,	36;10 36;30)	74r:4,115	Mod. Bastam, in NE Iran, NE of Damghān.
Bīza, see Pisa				
Bordeaux (text: Purdal)	35;15, (−0;34,	44;15 44;50)	74v:6,3	
Britain, Island of	9;0	50;30	74v:7,1	
Būdan, see Thule				
Bukhārā	97;30, (64;26,	39;10 39;47)	74r:5,15	In the Uzbekistan SSR.
Bulayda Sūdān	68;0,	17;0	72v:1,7	Presumably some locality in northern Sudan; given by no other source.
Būlghar country	90;0, (48;37,	50;30 54;19)	74v:7,9	Mod. Ulyanovsk, on the Volga downstream from Kazan (*Abū'l-F.*, transl., vol. 2, pt. 1, p. 326).
Bulūr	108;0,	36;0	74r:4,156	A district in the Khutlān region of NE Afghanistan (*LS*, p. 437).
Būna	38;0, (7;47,	33;50 36;55)	73v:4,17	Mod. Annaba, on the Mediterranean, NE corner of Algeria.
Bunduqiya, see Venice				
Burshān	50;0,	44;0	74v:6,7	*Abū'l-F.* (transl., vol. 2, pt. 1, p. 313) cites sources who claim this place was the capital of the Burjan(?) people, exterminated by the Germans, to the extent that no vestiges remain of the site.
Burūjird	84;30, (48;48,	34;20 33;55)	73v:4,96	Mod. Borujerd, in western Iran, south of Hamadān.
Būshang	95;40, (61;26,	37;50 34;20)	74r:4,136	Mod. Ghurian (*LS*, p. 431) in the NW corner of Afghanistan, west of Herat.
Bust	100;0, (64;21,	33;0 31;28)	73v:3,106	Or Bost, in southern Afghanistan, west of Kandahār and very close to Lashkar Gāh (*LS*, p. 344).
Būzjān	94;0, (60;36,	37;0 35;16)	74r:4,129	Mod. Torbat-e Jām (*LS*, p. 357) in NE Iran SSE of Mashhad.
Byzantium, see Constantinople				
Cadiz (Qādis) Island, middle of	18;14,	39;5	74r:5,3	The island on which modern Cadiz lies.

Place Name	Coordinates		Reference	Remarks
Cairo (text: Egypt, *Miṣr*)	63;0, (31;15,	30;20 30;3)	73r:3,30	
Canton, see Khanqū				
Caspian, see Khazar				
Ceuta, see Sibta				
Constantinople (text: Būzanṭiya, i.e. Byzantium, Qusṭanṭiniya)	59;50, (28;57,	45;0 41;2)	74v:6,9	
Cordova (text: Qurṭuba)	18;40, (−4;46,	35;40 37;53)	73v:4,3	
Crete (text: Aqrīṭish), Island	55;40, (25;8,	36;30 35;20)	73v:4,22	Mod. coordinates are for Iraklion.
Ctesiphon (text: Madā'in, Aywān Kisrā)	80;20, (44;36,	33;0 33;6)	73r:3,59	Near Baghdad.
Cyprus (text: Qubrus)	62;15,	35;15	73v:4,24	No city being named, no modern coordinates have been attempted.
Dabīl	80;30, (44;30,	38;0 39;45)	73v:4,79	In Armenia south of Yerevan (*LS*, p. 182).
Dahlak Island	71;0, (40;0,	14;0 15;30)	72v:1,8	Archipelago off Ethiopia.
Damār	77;15, (44;30,	13;30 14;33)	72v:1,19	Mod. Dhamar, in the Yemen south of Ṣan'ā.
Damascus	70;0, (36;19,	33;20 33;30)	73r:3,51	
Damāvand	86;0, (52;4,	36;15 35;47)	74r:4,107	In Iran, east of Tehran near the mountain of the same name. In most of the medieval tables written Dunbāwand.
Dāmghān	88;0, (54;22,	36;20 36;9)	74r:4,112	In northern Iran, east of Tehran.
Damietta (text: Dimyāṭ)	63;2, (31;48,	31;35 31;26)	73r:3,31	In Egypt.
Dandā[n]aqān	97;30, (61;45,	34;45 37;3)	74r:4,139	In the Turkmeniya SSR just SW of Mary (Marv, *LS*, p. 400).
Dāniya, see Denia				
Dara'a	21;6, (−11;3,	25;10 28;47)	72v:2,3	Mod. Dra, Morocco.

Place Name	Coordinates		Reference	Remarks
Dārābjird	90;15, (54;33,	28;15 28;45)	73r:3,91	Mod. Dārāb, in southern Iran, SE of Shīrāz.
Darqān	96;0, (62;10,	40;30 40;30)	74r:5,52	Mod. Darganata (*LS*, p. 451) on the Oxus (Amu Darya) in the Turkmeniya SSR.
Daybul	102;30, (67;45,	24;10 24;37)	72v:2,38	Near mod. Karachi, *LS*, p. 331.
Delhi (text: Dihlī)	114;18, (77;14,	28;13 28;40)	73v:3,107	
Denia (text: Dāniya)	29;10, (0;7,	39;6 38;51)	74r:5,7	On a peninsula protruding from the eastern coast of Spain.
Dīnawar	83;0, (47;32,	35;0 34;27)	73v:4,89	In southern Iran east of Kermānshāh (*LS*, p. 189).
Dongola, see Dunqula				
Dūba	43;50,	41;30	74r:5,19	According to the *Ḥudūd*, p. 321, this is the name of the Danube, but Kāshī's coordinates put it near Rome. No other source gives this locality.
Dunqula al-Ḥabasha	63;0, (30;27,	14;30 19;10)	72v:1,5	Mod. Dongola, Sudan.
Elat, see Ayla				
Fam al-Ṣulḥ	80;45, (45;51,	32;15 32;30)	73r:3,64	Near Kut in Iraq, cf. *LS*, pp. 28, 38.
Fārāb	98;30, (68;6,	45;0 42;48)	74v:6,20	Near mod. Timur (*LS*, p. 484), in the Kazakhstan SSR, north of Tashkent.
Faraḥ, see Madinat al-Faraj				
Farāwa	90;0, (56;23,	39;0 39;0)	74r:4,119	Mod. Kizil Arvat (*LS*, p. 472), in the Turkmeniya SSR, NW of Ashkhabad.
Farghāna, known as Andījān	102;0 (72;23,	42;20 40;48)	74v:5,71	Mod. Andizhan (*LS*, p. 477), in the Uzbekistan SSR.
Fāryāb	99;0, (64;55,	36;45 36;21)	74r:4,141	Mod. Khairābād in northern Afghanistan between Herat and Balkh (*LS*, p. 425).
Fās	18;15, (−5;0,	32;15 34;5)	73r:3,2	Mod. Fés (or Fez), Morocco.
Fayd	78;10, (42;34	26;50 27;8)	72v:2,27	In Najd of Arabia.

Place Name	Coordinates		Reference	Remarks
Fayyūm	63;15, (30;50,	29;15 29;19)	73r:3,32	In Egypt.
Fīrūzābād	87;30, (52;35,	28;10 28;51)	73r:3,85	In Iran, near Shīrāz.
Fura'a	77;30,	25;15	72v:2,26	In the Ḥijāz, according to *Abū'l-F.* (text, p. 94).
Furḍat al-Rūm	74;30,	46;50	74v:6,15	The name can be translated as "Byzantine Port." The coordinates are those given for Trebizond by other sources, but Kāshī locates it north of his Trebizond. No other source gives this name.
Galicia (text: [J]alīqiya)	20;0, (−5;45,	46;0 41;30)	74v:6,1	Mod. Zamora, in NW Spain. Other sources, for Samūra, give the same coordinates, hence the conclusion.
Genoa	41;0, (8;56,	41;20 44;24)	74r:5,18	
Ghāna, Gold Mine	39;0,	10;0	72v:0,3	Many other sources give this place, but with widely differing coordinates.
Gharnāṭa, see Granada				
Ghazna of Zābulistān	104;20, (68;28,	33;25 33;33)	73v:3,113	Mod. Ghazni, in east-central Afghanistan, SSW of Kābul.
Ghazza	66;10, (34;28,	32;0 31;30)	73r:3,35	Mod. Gaza, near the Mediterranean, SW of Jerusalem.
Ghudāmis	49;10, (9;30,	39;10 30;10)	73v:4,20	In western Libya.
Granada (text: Gharnāṭa)	21;40 (−3;35,	37;30 37;10)	73v:4,9	
Guadalajara, see Madīnat al-Faraj				
Gurganj, capital of Khwārazm	94;30, (59;10,	42;17 42;18)	77r:5,50	Mod. Kunya Urgench, cf. Jurjaniya. These coordinates are from the Aya Sofya MS; in the India Office copy Gurganj has been switched with Hazar Asp.
Hadia	66;0, (43;46,	7;0 7;21)	72v:0,11	Mod. Hado in Ethiopia.
Ḥadītha on the Euphrates	77;20, (42;22,	34;30 34;9)	73v:4,67	Mod. al-Ḥadītha, NW of Baghdad.
Ḥadītha on the Tigris	77;25, (42;49,	36;15 35;59)	73v:4,69	In northern Iraq, south of Mosul.
Ḥaḍra Island, see Algeciras				

Place Name	Coordinates		Reference	Remarks
Ḥakā(?) ibn Ya'qūb	77;20,	18;30	72v:1,21	Apparently in the Yemen, near Ṣa'ada.
Hajar of Baḥrayn	83;15,	25;15	72v:2,30	Near the coast of Arabia opposite the island of Baḥrayn; its port was modern al-'Uqayr (*Ḥud.*, p. 413).
Halāward	101;0,	37;30	74r:4,143	In the Tadjikistan SSR, on the Wakshāb (Vakhsh) east of Dushanbe (*LS*, p. 438).
Ḥalīqiya (for [J]aliqiya?) see Galicia				
Hamadān	83;0, (48;35,	35;10 34;46)	73v:4,90	In western Iran.
Ḥānsā, see Khānsā				
Ḥarrān	73;15, (39;1,	37;50 36;51)	73v:4,53	Mod. Atibasak, in southeastern Turkey south of Urfa.
Ḥasā	83;30,	24;15	72v:2,31	Presumably on the Arabian peninsula south of Baḥrayn.
Hatta[kh]	74;30,	37;45	73v:4,57	*Abū'l-F.* (transl., vol. 2, pt. 2) says that this is a castle in the region of Diyārbakr (in Turkey).
Haykal, see Port Vendres				
Hazār Asp	95;20, (61;5,	41;10 41;19)	74r:5,49	In the Uzbek SSR, east of Urgench. These coordinates are from the Aya Sofya MS; in the India Office copy Hazar Asp has been switched with Gurganj.
Heraqla	67;20, (31;26,	46;30 41;17)	74v:6,12	Mod. Ereğli, Zonguldak, in north-central Turkey, on the Black Sea.
Herāt	94;20, (62;10,	34;30 34;20)	74r:4,131	In the NW corner of Afghanistan.
Ḥijr (see also Ḥujr)	70;30, (40;40,	28;30 25;52)	73r;3,49	Mod. Ḥujr, in Arabia, north of Medina.
Ḥilla	79;15, (44;29,	32;15 32;28)	73r:3,53	In Iraq, south of Baghdad.
Ḥiṣn Dimlūh	74;40,	14,5	72v:1,12	In the Yemen, N. of Aden, according to *Abū'l-F.* (text, p. 90).
Ḥiṣn ibn 'Ammāra	90;0, (54;52,	30;20 26;34)	73r:3,90	Mod. Bandar-e Lingeh, on the Gulf coast of Iran, west of the Strait of Hormuz.
Ḥiṣn Mahdī	84;45,	30;45	73r:3,76	A site, presently unknown, in SW Iran, on the Kārūn below Ahvāz (*Abū'l-F.*, text, p. 316).

Place Name	Coordinates		Reference	Remarks
Ḥiṣn Manṣur	72;24, (38;15,	34;0 37;46)	73v:4,49	Mod. Adiyaman (*LS*, p. 123), in south-central Turkey, south of Malatiya.
Hīt	78;20, (42;50,	33;0 33;38)	73r:3,52	In central Iraq, on the Euphrates west of Baghdad.
Ḥomṣ	70;45, (36;43,	34;0 34;44)	73v:4,38	In north-central Syria.
Ḥujr(?)	81;10,	24;15	72v:2,28	Presumably on the Arabian peninsula near Baḥrayn.
Ḥulwān	82;15, (45;52,	34;0 34;28)	73r:3,68	Mod. Sar-e Pol-e Zahāb. (*LS*, p. 191), in western Iran between Kermānshāh and Qaṣr-i Shīrīn.
Hurmūz	92;0, (57;6,	25;0 27;7)	72v:2,36	In the Persian Gulf, cf. *LS*, pp. 318–321.
Ibiza (text: [Y]ābisa) Island	30;45,	38;30	73v:4,15	Off the Spanish coast, E of Alicante (*Abū'l-F.*, text, p. 190).
[Idfu?] (text: Anṣubā)	63;0, (32;52,	23;0 24;55)	72v:2,10	In Egypt between Luxor and Aswān. The restoration is drastic, but seems justified on the basis of other sources.
Īlāqī (or Īlāq)	99;10, (70;10,	43;20 41;1)	74v:5,57	Mod. Angren (from Persian Āhangarān) in the Uzbek SSR, east of Tashkent (*LS*, p. 482).
Indus (Mihrān) Source	126;0,	36;0	74r:4,157	
Irbīl	79;50, (44;1,	36;20 36;12)	73v:4,76	Mod. Arbil, in northern Iraq east of Mosul.
Isbānīkat	100;30, (68;30,	40;0 40;45)	74v:5,63	Sometimes called Banākat, in the Uzbek SSR, at the junction of the Syrdarya (Jaxartes) and the Angren (Īlāq). See *LS*, p. 482, and *Abū'l-F.*, text, p. 499.
Iṣfahān	86;40, (51;41,	32;25 32;41)	73r:3,80	Mod. Esfahān in central Iran.
Isfarā'in	91;40, (57;26,	37;55 37;3)	74r:4,121	Mod. Miānābād (*LS*, p. 393), in NE Iran between Mashhad and Gonbād-i Kāvūs.
Isfījāb of Shāsh	99;50, (69;5,	43;30 42;16)	74v:5,60	Mod. Sayram (*LS*, p. 483), in the Kazakhstan SSR, eight miles east of Chimkent.
Ishbīla, see Seville				
Ishbūna, see Lisbon				

Place Name	Coordinates		Reference	Remarks
Iskenderun (text: Bāb Iskandariya)	70;15, (36;8,	36;10 36;37)	73v:4,36	Turkish port on the Mediterranean.
Isnā	62;0, (32;30,	23;30 25;16)	72r:2,9	On the Nile between Aswān and Qūṣ.
Iṣṭakhr	88;30, (52;56,	30;0 30;6)	73r:3,88	Mod. Sivand, in Iran, near Shīrāz.
Jaen, see Jayyān				
Jaffa	66;15, (34;45,	32;20 32;3)	73r:3,37	
[J]alīqiya, see Galicia				
Jalūlā	81;10, (45;10,	33;30 34;16)	73r:3,66	In Iraq, near the Iranian border at Khāniqīn.
Jamkūt	176;0,	5;0	72v:0,23	In Sanskrit, Yamakoti. This is the Indian Kangdez (which see), and some sources place it on the equator in the extreme east (*India*, transl., vol. 1, p. 304).
[J]anad	75;30,	14;30	72v:1,15	In the Yemen, N of Ta'izz, according to *Abū'l-F.* (text, p. 90).
Jand	97;45,	43;30	74v:5,54	In the Kazakhstan SSR, on the Syrdarya (Jaxartes) between Tashkent and the Aral (*LS*, p. 486).
Jannāba	87;20, (50;33,	30;15 29;34)	73r:3,83	Mod. Ganaveh (*LS*, p. 294), on the Gulf coast, west of Shīrāz.
Jarjariya	80;30, (45;5,	33;15 32;46)	73r:3,63	Near mod. Zubaydiyah, on the Tigris south of Baghdad.
[J]armī, capital of Ḥabasha (Abyssinia)	65;0,	9;30	72v:0,9	Ḥarmī in text.
Jayḥūn (Oxus, Amū Daryā) Source, middle of the lake.	110;0,	48;0	74v:7,14	The actual source of the Oxus is far SW of the location given here.
Jayrūn (?)	66;30,	35;15	73v:4,27	No other source mentions this place, the coordinates seemingly putting it in the Mediterranean south of Athens.
[J]ayyān	21;40, (−3;48,	38;50 37;46)	73v:4,10	Mod. Jaen, in southern Spain, ESE of Cordova.
Jazīra ibn 'Amrū	75;30, (42;11,	37;30 37;21)	73v:4,59	Mod. Cizre, on the Tigris at the eastern end of the Turkish-Syrian border.

Place Name	Coordinates		Reference	Remarks
Jerusalem (text: Bayt al-Muqaddis)	66;30, (35;13,	31;50 31;47)	73r:3,38	
Jidda	76;0, (39;10,	21;0 21;30)	72v:2,23	On the Red Sea west of Mecca.
Jirba Island	19;15, (11;0,	32;15 33;39)	73r:3,3	Mod. Jerba, Tunisia.
Jīruft	93;0, (57;48,	27;30 28;41)	72v:2,37	In Fārs, Iran.
Jubla	75;0, (44;46,	13;30 14;55)	72v:1,14	Mod. Jubil in the Yemen.
Juḥfa	74;0,	22;0	72v:2,20.	In the Ḥijāz.
Jurash	77;50,	17;15	72r:1,25	In the Yemen near Najrān, according to *Abū'l-F.* (text, p. 94).
Jurbādaqān	85;35, (50;18,	34;15 33;23)	74r:4,105	Mod. Golpāyegān, in Iran between Hamadān and Esfahān.
Jurjān	90;0, (55;11,	36;50 37;15)	74r:4,118	Mod. Gonbad-i Kāvūs (*LS*, pp. 8, 377), NE Iran.
Jurjāniya, Khwārazm	94;0, (59;10,	42;45 42;18)	74r:5,48	Mod. Kunya-Urgench (*LS*, p. 441), in the Uzbekistan SSR south of the Aral. We give the same modern coordinates for Gurganj. For many reasons, including shifts in the bed of the Oxus, the situation concerning these two place names is complicated.
[J]ymy(?) on the Nile	63;15,	9;0	72v:0,7	Text has Ḥīmī.
Kābul	104;40, (69;10,	34;30 34;30)	74r:4,152	Capital of Afghanistan.
Kafā	67;14, (35;23,	48;0 45;3)	74v:7,6	Mod. Feodosiya on the Crimean peninsula.
Kairouan (text: Qayrawān)	41;15, (10;1,	31;15 35;42)	73r:3,15	In Tunisia.
Kajār and Kalār	86;50, (51;21,	36;25 36;29)	74r:4,109	Mod. Kalārdasht (*LS*, p. 373), near the Caspian coast of Iran, north of Tehran.
Kalah(?) Island	140;0,	8;0	72v:0,21	In the Indian Ocean.
Kalār, see Kajār and Kalār				
Kamrun, see Qāmrūn				
Kanbāyat	109;20, (72;35,	26;20 22;20)	73r:2,43	Mod. Khambhat, or Cambay, in Gujerat, India, at the head of the Cambay Gulf (*Maq.*, p. 156).

Place Name	Coordinates		Reference	Remarks
Kangdez	180;0	0;0	72v:0,25	A mythical Iranian castle, built by the earliest of the kings. Kāshī follows tradition by placing it at the extreme east of the inhabited portion of the globe.
Kanoj	114;50, (79;56,	26;35 27;2)	73r:2,44	Mod. Kannauj, on the Ganges.
Karaj	84;45, (50;58,	34;0 35;48)	74r:4,99	In north-central Iran, between Tehran and Qazvīn.
Karak	67;30, (35;42,	31;30 31;11)	73r:3,43	In southern Jordan.
Karakorum (text: Qaraqurūm)	115;0, (102;50,	46;0 47;10)	74v:6,29	In Mongolia, west of Ulan Bator.
Karsh (?)	87;0,	47;50	74v:6,18	No other source mentions this locality, its coordinates seemingly placing it north of the Caspian.
Kāsān	101;35, (71;31,	42;0 41;14)	74v:5,69	Mod. Kasansay (*LS*, p. 480), in Uzbekistan SSR, north of Namangan
Kāshān	86;0, (51;35,	34;0 33;59)	74r:4,106	In Iran between Tehran and Esfahan.
Kāshghar	106;30, (76;2,	44;0 39;29)	74v:6,26	Mod. Kashi, Sinjiang Uygur Zizhiqu, China.
Kashmīr	108;40, (74;54,	35;0 32;43)	74r:4,155	Mod. coordinates are those of the present capital.
Kaslūna(?)	65;30,	46;20	74v:6,11	No other source mentions this place, whose coordinates put it between Kashī's Sinope and Heraqla.
Kāth	95;0, (59;53,	41;36 42;20)	74r:5,47	Mod. Kungrad (?), (*LS*, p. 446) in the Uzbekistan SSR. This location is a conjecture. Cf. also Gurganj and Jurjāniya. The Aya Sofya MS gives the longitude as 94;0°.
Kaulam	120;45, (76;38,	13;30 8;53)	72v:1,31	Mod. Quilon, Kerala, India.
Kawtam, or Kūtam	84;40, (49;58,	37;20 36;17)	73v:4,98	In northern Iran, near the modern Āstāneh east of Rasht.
Kāzirūn	87;0, (51;40,	29;15 29;35)	73r:3,81	In Iran, near Shīrāz.
Kerker, see Qirqir				
Khabīṣ	93;0, (57;44,	31;0 30;27)	73r:3,98	Mod. Shahdāb (*LS*, p. 308), in SE Iran just east of Kermān.

Place Name	Coordinates		Reference	Remarks
Khājū	123;30	42;15	74v:5,76	*Abū'l-F.* (transl., vol. 2, pt. 2, p. 125) says this place is in northern China. Minorski (*Ḥud.*, p. 233) makes it Kua-chou, the An-hsi oasis on the Su-lo-ho River.
Khāltān of Makrān	99;0,	30;0	73r:3,104	No other source mentions this locality, in SE Iran.
Khānbāliq	124;0, (116;26,	46;0 39;55)	74v:6,30	A footnote in *Abū'l-F.* (transl., vol. 2, pt. 2, p. 230) says this is Peking (present Beijing).
Khānjū, China	162;30, (118;36,	13;0 24;53)	72v:1,36	Mod. Quazhou
Khānqū, China	160;0, (113;20,	14;0 23;8)	72v:1,35	Mod. Guangzhou (Canton, *Ḥud.*, p. 227, *Abū'l-F.*, transl., vol. 2, pt. 2, p. 122). Perhaps the correct Arabic form was Khānfū, but all five sources give the *qāf*.
[Kh]ānsā (?Ḥānsā?)	115;15,	28;30	73v:108	According to *Abū'l-F.* (transl., vol. 2, pt. 2, p. 124) a Chinese seaport. There is confusion between this name and Khānfū, Khānqū.
Khaybar	75;20, (39;12,	24;20 25;48)	72v:2,22	In Arabia north of Medina.
Khaywān	77;21,	15;20	72v:1,23	According to *Abū'l-F.* (transl., vol. 2, pt. 1, pp. 120, 128, 129) a village in southern Yemen.
Khazar (Caspian) Sea, end of	89;0,	50;0	74v:7,13	
Khuḍrā' Island, see Algeciras				
Khujand	100;35, (68;40,	41;55 40;14)	74v:5,64	Mod. Leninabad (*LS*, p. 462), in the Tadzhikistan SSR, south of Tashkent.
Khurram	105;20,	36;0	74r:4,154	No other source mentions this place, which by its coordinates would be in eastern Afghanistan.
Khutan	107;0, (79;57,	42;0 37;7)	74v:5,73	Mod. Hotan (*LS*, p. 489), in the Uighur autonomous region of China.
Khuwār	87;10, (52;20,	35;40 35;15)	74r:4,110	Mod. Garmsar (*LS*, p. 367), in north-central Iran, SE of Tehran.
Khvoy	79;40, (45;2,	37;40 38;32)	73v:4,74	In the extreme NW corner of Iran.
Khwāftand	100;50, (70;55,	42;50 40;33)	74v:5,65	Mod. Kokand (*Ḥud.*, p. 355, *Abū'l-F.* text, p. 499) in Central Asia.

Place Name	Coordinates		Reference	Remarks
Khwārazm, see Gurganj, also Jurjāniya, also Kāth				
Khwāsh	95;40, (62;49,	33;0 31;29)	73r:3,101	Mod. Khāsh, Afghanistan.
Kīj of Makrān	99;0,	28;30	73r:3,103	Now a region in west Pakistan, variously called Kīz and Kech (*Maq.*, p. 157).
Kirmān	91;30, (57;5,	30;5 30;18)	73r:3,94	Mod. Kerman, in SE Iran.
Kish, see Shahr Sabz				
Konya (text: Qūniya)	66;30, (32;30,	41;40 37;51)	74r:5,27	In south-central Turkey, SW of the Tuz Gölü.
Kūfa	79;30, (44;25,	31;30 32;2)	73r:3,54	South of Baghdad on the Euphrates.
Kūkūh	54;10,	10;0	72v:0,4	Bīrūnī (*Abū'l-F.*, text, p. 156) says this is in Africa, E of Ghāna.
Kūrī, Lake, middle of	63,	0;0	72v:0,6	No other source reports this place presumably in central Africa.
Lahore (text: Lahāwur)	109;20, (74;22,	31;50 31;34)	73v:3,118	
Lake, see Kūrī				
Lāmrī Island	137;0,	9;0	72v:0,20	*Abū'l-F.* (transl., vol. 2, pt. 2, p. 131) states that this is in the Indian Ocean.
Lamṭa, called Nasawā(?)	17;30,	24;0	72v:2	In the western Maghrib.
Lamuryā (Morea), city on an island	55;14,	43;15	74r:5,20	Morea was the medieval name for the Peloponnesian peninsula. In *Abu'l-F.* (transl., vol. 2, pt. 1, p. 275) it is stated that to the name the Romance article La was added. But the latitude given by all sources (*Tus.*, *Ulg.*, *IO*) is impossible for this.
Latakia (text: Lādhiqiya)	70;40, (35;47,	35;15 35;31)	73v:4,37	Syrian seaport on the Mediterranean.
Lisbon (text: Ishbūna)	16;15, (−9;8,	42;40 38;44)	74r:5,1	
Lunbardiya, see Milan				
Luxor (text: Uqṣur)	61;30, (32;24,	24;15 25;41)	72v:2,8	On the Nile, between Aswān and Asyūt.

Place Name	Coordinates		Reference	Remarks
Maʿabar	102;0,	17;30	72v:1,30	Kāshī's longitude is probably a corruption of *Abū'l-F.*'s (vol. 2, pt. 2, p. 121) 142°, who puts this place on the Coromandel coast of SE India.
Maʿarat al-Nuʿmān	71;35, (36;41,	45;0 35;37)	73v:4,44	In northwestern Syria.
Macedonia	65;0,	41;1	74r:5,25	
Madāʾin, see Ctesiphon				
Maʿdan Dhahab (Gold Mine)	67;35,	21;45	72v:2,14	In southern Egypt?
Maʿdan Zumurrud (Emerald Mine)	66;0,	21;0	72v:2,12	In the Sudan?
Madina (Medina)	75;20, (39;35,	24;50 24;30)	72v:2,21	
Madina Sālim, see Medinaceli				
Madīna Walīd, see Valladolid				
Madīnat al-Fara[j]	25;15, (−3;10,	36;40 40;37)	73v:4,12	Mod. Guadalajara, in Spain NE of Madrid. The name is from Wādī al-Ḥijara, which is used by most of the medieval sources. But *Abū'l-F.* (transl., vol. 2, pt. 1, p. 255) cites the appellation used by Kāshī.
Mahdiya	42;15, (11;3,	32;30 35;29)	73r:3,16	In Tunisia.
Mahra	85;0,	16;0	72v:1,27	A region, the easternmost state of South Yemen.
Mahrāj Island	150;0,	1;0	72v:0,22	In the Indian Ocean. Other sources give the same coordinates for Sarīra Island except the latitude is south of the equator.
Mahūra, the city of Brahma	116;0, (77;42,	24;40 27;30)	73r:2,45	Mod. Mathura, in India.
Majorca (text: Marqa) Island	34;7,	38;30	73v:4,16	Off the east coast of Spain.
Mājūj (Magog)	172;30,	63;0	74v:8,9	This word stems from the mythical Gog and Magog mentioned in both the Bible and the Koran. They were supposed to be peoples living somewhere in the far northeast (*EI*, vol. 4, p. 1142)

Place Name	Coordinates		Reference	Remarks
Makrān, see Tīz				
Málaga (text: Mālaqa)	26;0, (−4;24	37;0 36;43)	73v:4,13	In southern Spain, on the Mediterranean coast.
Malaṭiya	71;0, (38;18,	37;0 38;22)	73v:4,41	In central Turkey NW of Diyarbakr.
Malāzjird	75;0, (42;30,	39;30 39;9)	74r:5,33	Mod. Malazgirt, in east-central Turkey, north of Lake Van.
Mālīn	95;40, (62;13,	34;30 34;12)	74r:4,135	Mod. Rauzabāgh (*LS*, p. 407), in the NW corner of Afghanistan, just south of Herāt.
Māliq	102;30,	44;0	74v:6,23	According to Kāshī's coordinates, just west of Uzkand, which see.
Manbij	72;50, (37;55,	36;30 36;32)	73v:4,51	Mod. Membij, in NW Syria near the Turkish border.
Mānjū (Manchu ?) of China	126;0,	39;0	73r:2,48	
Manṣūra of Sind	105;0, (68;47,	26;40 25;53)	73r:2,41	In India, forty-seven miles NE of Hyderabad, Sind (*Maq.*, p. 93).
Marāgha	82;0, (46;13,	37;20 37;25)	73v:4,84	In NW Iran, south of Tabrīz.
Marand	80;0 (45;0,	37;50 38;25)	37v:4,77	In the NW corner of Iran, NW of Tabrīz.
Marbella	24;40, (−4;53,	35;50 36;31)	73v:4,11	In southern Spain, on the coast, between Gibraltar and Málaga.
Mārdīn	74;15, (40;43,	37;50 37;19)	73v:4,55	In southeastern Turkey, near the Syrian border.
Mārī Kirmān	65;45,	50;40	74v:8,2	No other source gives this place, which, according to Kāshī's coordinates, is just NE of Aqjakirmān, which see.
Mārib, see Sabā				
Mārida, see Merida				
Mārqa, see Majorca Island				
Marrākish	21;15, (−8;0,	29;15 31;49)	73r:3,5	Mod. Marrakech, Morocco.
Marv-i Rūd, known as Murghāb	94;40, (63;20,	34;30 35;34)	74r:4,134	Mod. Bālā Murghāb (*LS*, p. 404) in the NW corner of Afghanistan, near the Soviet border.
Marv-i Shāhijān	94;0, (61;54,	37;40 37;42)	74r:4,130	This is Great Marv (Merv) mod. Mary, in the Turkmeniya SSR (*LS*, p. 397).
Maṣīṣa	69;15, (35;35,	36;45 36;57)	73v:4,32	Mod. Misis (*LS*, p. 130), in south-central Turkey east of Adana.

Place Name	Coordinates		Reference	Remarks
Maymand	101;55, (65;5,	33;20 31;45)	73v:3,112	In Afghanistan, between Kandahar and Girishk (*LS*, p. 425).
Mayyāfāriqīn	74;15, (41;0,	38;0 38;8)	73v:4,56	Mod. Silvan (*LS*, p. 111) in southern Turkey, east of Diyarbakr.
Mecca	77;10, (39;49,	21;40 21;26)	72v:2,24	
Medinaceli (text: Madīna Sālim)	28;15, (−2;26,	43;15 41;10)	74r:5,6	In central Spain, NE of Guadalajara.
Menorca Island (text. Minurqa)	34;10, (4;15,	39;40 39;54)	74r:5,11	One of the Balearic Islands, in the Mediterranean east of Spain. The modern coordinates are those of Mahon.
Merida (text: Mārida)	20;15, (−6;20,	38;15 38;55)	73v:4,7	In SW Spain, east of Badajoz.
Mīāneh	82;30, (47;45,	37;0 37;23)	73v:4,87	In NW Iran, SE of Tabrīz.
Mihrān River, see Indus				
Milan (text: [L]unbardia, for Lombardy)	40;30, (9;12,	43;50 45;28)	74v:6,4	
Mirbāṭ	82;0, (54;42,	12;0 16;58)	72v:0,17	In Oman, a port east of Salālah.
Misīla	38;40, (4;31,	30;20 35;40)	73r:3,13	Mod. M'Sila, in Algeria SE of Algiers.
Morea, see Lamuryā				
Mosul	77;0, (43;8,	36;50 36;21)	73v:4,65	In north-central Iraq.
Muʿjam(?)	74;15,	17;15	72v:1,10	No other source gives this locality, which seems to be in Arabia, SW of Najrān.
Mūltān	106;25, (71;36;	29;40 30;10)	73v:3,115	In Pakistan.
Muqadīshū	72;0, (45;21,	2;0 2;2)	72v:0,13	Mod. Mogadiscio in Somalia.
Mūqān	83;0,	38;0	73v:4,92	A steppe region bordering on the SW coast of the Caspian south of Bākū. Kāshī doubtless intends the chief city (*LS*, p. 175).
Murcia (text: Murtasiya)	22;50, (−1;8,	39;20 37;59)	74r:5,5	In SE Spain, north of Cartagena.
Murtasiya, see Murcia				
Nabalūna, see Pampluna				

Place Name	Coordinates	Reference	Remarks
Nabghīkand	97;30, 46;40	74v:6,19	No other source gives this place, which seems to be NW of Fārāb, which see.
Nahāvand	83;45, 34;20 (48;21, 34;13)	73v:4,94	In western Iran, south of Hamadān.
Nahlawāra	108;20, 28;30 (72;11, 23;51)	73v:3,117	In India. We identify it with Nahrawāra-Patan in Gujerat (*Maq.*, p. 158).
Nahr al-Malik	80;50, 33;25	73r:3,65	A town named for a famous canal, near Baghdad (*LS*, p. 68).
Najrān	77;0, 19;15 (44;19, 17;31)	72v:1,17	In Saudi Arabia near the border with Yemen.
Nakhjavān	81;45, 37;45 (45;24, 39;12)	73v:4,81	Mod. Nakhichevan, in the ASSR of the same name, in the Caucasus.
Nakhshab, called Nasaf, or Qarshī	98;0, 39;0 (65;45, 38;53)	74v:5,55	Mod. Karshi (*LS*, p. 470), in the Uzbekistan SSR, SW of Samarqand.
Narbonne (text: Arbūna)	36;15, 43;20 (3;0, 43;11)	74r:5,14	In southern France, near the Mediterranean, and the Spanish border.
Nasaf, see Nakhshab			
Nasawā, see Lamṭa			
Nawbandagān	87;15, 30;10 (51;30, 30;14)	73r:3,82	Mod. Nūrābād (*LS*, pp. 263–265), in SW Iran, NW of Shīrāz.
Nawshahr	78;20, 38;10	73v:4,70	According to Kāshī's coordinates, this place is between Diyarbakr and Khoi, i.e. near the modern Irano-Turkish border. No other source cites it.
Niqarnīt(?) Island	58;50, 42;15	74r:5,21	Kāshī's coordinates may put this in the Sea of Marmora. *Khwar.* has almost the same coordinates for an Alus peninsula (Gr. Elaious).
[N]īrūn (text; Bīrūn)	104;30, 24;45 (68;24, 25;23)	72v:2,40	Mod. Hyderabad in Sind of Pakistan, on the Indus near its mouth (*Maq.*, p. 98).
Nīshāpūr	92;30, 36;21 (58;49, 36;13)	74r:4,123	Mod. Neyshābūr, in NE Iran, west of Mashhad.
Niṣībīn	75;20, 36;40 (41;11, 37;5)	73v:4,60	Mod. Nusaybin, in SE Turkey, on the Syrian border.
Nuʿmāniya	80;20, 33;15 (45;23, 32;34)	73r:3,61	Mod. An-Nuʿmānīyah, on the Tigris south of Baghdad.

Place Name	Coordinates		Reference	Remarks
Nūqān	92;45, (59;31,	38;40 36;30)	74r:4,126	Mod. Ṭūs (*LS*, p. 388), in NE Iran near Mashhad.
Ocean, beside the	20;0,	16;0	72v:1,1	The Atlantic.
Ocean Coast	11;0,	0;0	72v:0,1	Eastern edge of the Atlantic.
Ocean Sea, beside the	53;0,	61;0	74v:8,1	
Ocean Sea, end of	—	71;0	74v:8,8	
Palermo, capital of Sicily	45;15, (13;23,	37;10 38;8)	73v:4,19	
Pamplona (text: Nabalūna)	34;15, (−1;39,	45;15 42;49)	74v:6,2	In NE Spain, ESE of Bilbao.
Peking, see Khānbāliq				
Peloponnesus, see Lamuryā				
Pisa (text: [Bīz]a)	42;0, (10;24,	47;0 43;43)	74v:6,6	
Port Vendres (text: Haykal)	34;15, (3;6,	43;15 42;31)	74r:5,12	In France, on the Mediterranean, near the Spanish border.
Purdal, see Bordeaux				
Qabāliq(?)	108;0,	44;0	74v:6,27	The late Persian zījes give this locality, presumably in western Sinkiang (Xinjiang).
Qādis Island, see Cadiz				
Qā'in	93;20, (59;6,	36;30 33;43)	74r:4,127	Mod. Qāyen, in eastern Iran north of Bīrjand.
Qālīqalā	73;15,	39;0	73v:4,54	*LS*, p. 117, identifies this with Erzerum,. for which Kāshī has separate coordinates. *LS* also says Qālīqalā is the district north of Erzerum.
Qāmrūn Mountains	135;0,	10;0	72v:0,19	Probably the ranges (mod. Kamrup) in Bhutan, north of the Goalpara and Kamrup districts of Assam (*Maq.*, p. 160).
Qanbala(?) Island	21;0,	3;0	72v:0,2	In the four other sources giving this place the coordinates are 52;0, −3;0, i.e. south of the equator, and some give it as the capital of Zanj.
Qandahār	107;40, (65;47,	33;0 31;36)	73v:3,116	In south-central Afghanistan.
Qaraqurūm, see Karakorum				

Place Name	Coordinates		Reference	Remarks
Qarīnayn	97;15,	37;15	74r:4,138	According to *LS*, p. 400, this place is in the Turkmeniya SSR on the Murghāb near Mary (Merv).
Qarqisiyā	74;40, (40;21,	36;40 35;11)	73v:4,58	Mod. Qataʿa, at the junction of the Euphrates with the Khabur.
Qarshī, see Nakhshab				
Qaṣr ʿAbd al-Karīm	18;30, (−5;56,	37;40 35;4)	73v:4,2	Mod. Ksar-el-kebir, in northern Morocco, south of Tangier.
Qaṣr Aḥmad	51;25,	33;30	73r:3,22	In the Maghrib near Barca.
Qaṣr ibn Hubayra	80;30, (44;42,	32;45 32;22)	73r:3,62	Mod. Hāshimiyah, in central Iraq south of al-Ḥillah.
Qaṣr Shīrīn	81;40, (45;36,	33;45 34;32)	73v:4,82	In western Iran on the Iraqi border, near Kermānshāh.
Qaṭīf	74;0, (50;0,	25;0 26;30)	72v:2,18	In Baḥrayn.
Qayrawān, see Kairouan				
Qaysāriya Rūm	67;15, (35;28,	40;40 38;42)	74r:5,28	Mod. Kayseri, in central Turkey.
Qaysāriya Shām	66;30, (35;42,	32;50 33;14)	73r:3,41	Mod. Baniyas, anc. Caesarea Philippi, in SW Syria.
Qazvīn	85;0, (50;0,	36;15 36;16)	74r:4,101	In NW Iran, south of Rasht.
Qinnisrīn	72;0, (37;10,	35;30 35;57)	73v:4,46	In NW Syria due south of Aleppo.
Qirmisīn	83;0, (47;4,	34;32 34;19)	73v:4,88	The modern coordinates given are those of Kermanshah (following *LS*, p. 186) in west-central Iran.
Qirqir	65;30,	50;0	74v:7,5	According to *Abū'l-F.* (transl., vol. 2, pt. 1, p. 319) this place is in the interior of the Crimea north of Sarikirmān, which we have failed to locate.
Qolzum on the (Red) Sea	64;15, (32;33,	28;30 29;59)	73r:3,33	Near Suez, ancient Klysma, mod. coordinates from *Bat.* vol. 2, p. 52, no. 252.
Quba	101;50, (72;5,	42;50 40;34)	74v:5,70	Mod. Kuva in the Uzbekistan SSR south of Andizhan (*LS*. pp. 478, *Ḥudūd*, p. 355).
Qubrus, see Cyprus				
Qum	85;40, (50;57,	34;45 34;39)	74r:4,104	Mod. Qom, in central Iran, SSW of Tehran.

Place Name	Coordinates		Reference	Remarks
Qūnī(?) Islands	——,	65;0	74v:8,7	Some sources, following Ptolemy, give coordinates of (30, 63) to the island of Thule.
Qūniya, see Konya				
Qurṭuba, see Cordova				
Qūṣ	61;30, (32;48,	24;35 25;53)	72v:2,6	In the Ṣa'īd of Egypt.
Quẓayr(?)	69;0,	26;0	72v:2,17	In Arabia?
Rām Hurmuz	85;45, (49;38,	31;0 31;15)	73r:3,79	In SW Iran, east of Ahvāz.
Ramla	66;30, (34;52,	32;10 31;56)	73r:3,40	In Palestine.
Raqqa	73;0, (39;3,	36;15 35;57)	73v:4,52	In north-central Syria, on the Euphrates.
Rashīd	62;4, (30;25,	31;30 31;25)	73r:3,29	At one of the mouths of the Nile, NE of Alexandria.
Rayy	86;20, (51;27,	35;0 35;35)	74r:4,108	In north-central Iran, just south of Tehran.
Rhodes, Island	61;40, (28;16,	36;15 36;25)	73v:4,23	
Ribāt-i Amīr	105;0,	34;0	73v:3,114	According to these coordinates, this place is in the vicinity of Kābūl, but Ulugh Beg says it is in Makrān. Bīrūnī (*India*, transl., vol. 1, p. 317) calls it also Kandī.
Roman (Mediterranean) Sea, edge of	25;0,	32;0	73r:3,8	
Rome (text: Rumiya Kubrā)	[4]5;0 (12;30,	41;50 41;53)	74r:5,44	
Rukhkhaj	103;15,	32;50	73v:3,110	The region in Afghanistan watered by the Qandahār River (*LS*, p. 339).
Rūs	102;20,	43;20	74v:5,72	Kāshī's (and Ulugh Beg's) coordinates put this in the region of Uzkand (Uzgen). There are many references to the Rūs (future Russians) in *Ḥudūd*.
Ṣa'ada	77;20, (43;45	17;15 17;0)	72v:1,22	Mod. Sadah, in the Yemen.
Sabā, also called Mārib	73;0, (45;30,	14;0 15;30)	72v:1,9	In Yemen, the site of the famous ancient dam.
Sabzavār	91;30, (57;38,	36;0 36;13)	74r:4,120	In NE Iran, west of Mashhad.

Place Name	Coordinates		Reference	Remarks
Sadubān (?)	104;15,	28;15	73v:3,111	No other source gives this place. It may be a corruption of Sadusān, mod. Sehwan, in Sind of Pakistan, with modern coordinates of (67;52°, 26;26°).
Ṣaghaniyān	102;40, (67;58,	38;50 38;27)	74r:4,148	Mod. Saryassiya, in the Uzbekistan SSR north of Termez.
Salā (?) Heights of China	180;0,	5;0	72v:0,24	
Salmās	79;15, (44;50,	37;40 38;13)	73v:4,73	In the NW corner of Iran west of Lake Urmiya.
Samanjān	102;0, (68;3,	37;15 36;15)	74r:4,146	Mod. Samangān, northern Afghanistan, SE of Mazār-i Sharīf.
Samarqand	99;0, (66;57,	40;0 39;40)	74v:5,56	In the Uzbekistan SSR.
Sāmarrā	79;15, (43;52,	34;15 34;13)	73v:4,72	In Iraq, north of Baghdad.
Samnān	87;20, (53;25,	36;40 35;30)	74r:4,113	Mod. Semnān, in north-central Iran east of Tehran.
Samos, Island	52;40, (26;59,	38;10 37;44)	73v:4,21	Mod. coordinates are for Vathi.
Sāmsūn	69;20, (36;22,	46;40 41;17)	74v:6,14	On the Black Sea coast of Turkey.
Ṣanʿā, capital of Yemen	77;0, (44;14,	14;30 14;23)	72v:1,18	
Santarīn	18;10, (−8;40,	42;35 39;14)	74r:5,2	Mod. Santarem in Portugal, NNE of Lisbon.
Santiago (text: Shantiyāqū	19;0, (−8;33,	49;0 42;52)	74v:7,2	Mod. Santiago de Compostella, in the NW corner of Spain.
Ṣaqāliba (Slavs), the ignorant	——,	64;0	74v:8,6	Other sources give a longitude of 20;30° to this entry, which would put it in NW Europe.
Ṣaqjū (?)	58;30, (28;9,	50;0 45;15)	74v:7,3	Mod. Isaccea in Romania, west of Tulce on the Danube delta (*Abū'l-F.*, transl., p. 316).
Sarakhs	94;30, (61;7,	36;0 36;32)	74r:4,132	At the NE corner of Iran, on the border.
Sarāndīb Island	130;0,	12;0	72v:0,18	Mod. Ceylon.
Saraqusṭa	31;30, (−0;54,	42;30 41;39)	74r:5,9	Mod. Zaragoza, in northeastern Spain.
Ṣarāy	86;0, (45;55,	48;0 51;30)	74v:7,11	Mod. Saratov (?), in the USSR, on the Volga, NE of Volgograd.

Place Name	Coordinates		Reference	Remarks
Sardinia, Island	41;15, (9;8,	38;15 39;13)	73v:4,18	The modern coordinates given are for Cagliari.
Sārī	88;0, (53;6,	37;0 36;33)	74r:4,114	On the Caspian coastal plain of Iran ENE of Tehran (*LS*, pp. 370–375).
Sarīr Alān	83;0,	44;0	74v:6,16	These two words are the names of a pair of tribes domiciled in Daghistān, in the Caucasus (*Abū'l-F.*, transl., vol. 2, pt. 2, p. 155; *Ḥud.*, p. 447).
Sarīra, see Mahrāj				
Sarmīn (?)	71;50, (36;40,	35;45 35;55)	73v:4,45	In Syria between Aleppo and Maʿarat al-Nuʿmān (*Abū'l-F.*, transl., vol. 2, pt. 2, p. 42).
Sarūj	72;40, (38;59,	34;50 36;41)	73v:4,50	Near Tell el-Abyaḍ. in north-central Syria, on the Turkish border.
Sāveh	85;0, (50;22,	36;0 35;0)	74r:4,100	In NW Iran, SW of Tehran.
Sawākin Island	48;30, (37;17,	17;0 19;8)	72v:1,3	The modern Suakin, in Sudan, is on the Red Sea coast. Kāshī's longitude is badly off (*Abū'l-F.*, vol. 2, pt. 2, p. 128).
Ṣaymara	81;50, (47;7,	34;40 32;20)	73v:4;83	In western Iran, west of Khurramābād (*LS*, p. 202).
Seville (text: Ishbīla)	18;50, (−5;59,	36;15 37;24)	73v:4,4	
Shahr Sabz, or Kish	99;30, (66;49,	39;30 39;5)	74v:5,58	Mod. Shakhrisyabz, in the Uzbekistan SSR, south of Samarqand.
Shahrazūr	80;20, (46;9,	35;30 35;30)	73v:4,78	Mod. Dezh Shāhpūr (*LS*, p. 190), in NW Iran, west of Sanandaj.
Shahrūd, see Suhraward				
Shal[j]	100;30,	44;0	74v:6,22	According to Kāshī's coordinates (and *Abū'l-F.*, vol. 2, pt. 2, p. 224) this place is near Ṭarāz, (which see) hence in the Kazakhstan SSR.
Shamākhī, capital of Shirvān	84;30, (48;37,	40;50 40;38)	74r:5,43	Mod. Shemakha, in the Azerbaydzhan SSR, west of Baku.
Shamkūra	83;0, (46;0,	41;50 40;50	74r:5,39	Mod. Shamkhor, in the Azerbaydzhan SSR, NW of Kirovabad (*LS*, p. 178).
Sharja	74;40,	17;50	72v:1,13	According to *Abū'l-F.* (transl., vol. 2, pt. 1, p. 122), this place is in the Yemen, hence distinct from Sharja in the Gulf.

Place Name	Coordinates		Reference	Remarks
Shāsh, see also Binkath	109;0,	42;30	74v:5,74	Shāsh was a region of Central Asia, its capital Binkath, which Kāshī also gives. These coordinates, cited by other sources, make little sense.
Shāwkath	100;30,	41;0	74v:5,62	According to *Abū'l-F.* (transl., vol. 2, pt. 2, p. 225) this place is in the region of Shāsh, hence the modern Uzbekistan SSR.
Shayzar	71;10, (36;34,	34;50 35;20)	73v:4,42	Near mod. Suran in Syria, NW of Ḥamā.
Shibām, Ḥaḍramaut	81;15, (48;34,	12;30 15;58)	72v:0,16	In southern Yemen, north of al-Mukalla.
Shimshāṭ	73;15, (39;26,	40;0 38;43)	74r:5,32	In east-central Turkey; modern coordinates have been interpolated from the remarks in *LS*, p. 116.
Shīrāz	88;0, (52;34,	29;30 29;38)	73r:3,86	In south-central Iran.
Shirvān, see Shamākhī				
Shu[gh]r (and) Bakas	71;0, (36;20,	35;30 35;52)	73v:4,39	Twin castles, once Crusader, halfway between Antioch and Apamia, mod. Famiyah, (*Guide Bleu*, p. 476).
Sibta	19;15, (−5;19,	35;30 35;53)	73v:4,5	Mod. Ceuta, in Morocco opposite Gibraltar.
Sicily, a big island	65;0,	36;0	73v:4,25	The longitude is thirty degrees too large, putting the island in the eastern Mediterranean, but other sources give the same value. See also Palermo. Presumably Kāshī drew on at least two different sources.
Sidon (text: Ṣayda)	68;30, (35;22,	33;0 33;32)	73r:3,48	
Sijilmāsa	20;0,	31;30	73r:3,4	In the Maghrib.
Sijistān, see Zaranj				
Siktāsh (?)	130;0,	39;10	74v:5,78	Kāshī's coordinates put this place east of Khānbāliq (Peking?).
Sindān	115;20,	19;15	72v;1,34	Mod. Sanjān, in India north of Bombay (*Maq.*, p. 102).
Sīnīz	85;30,	30;0	73r:3,78	On the Persian Gulf.
Sinjār	76;0, (41;51,	36;0 36;20)	73r:4,62	In NW Iraq, west of Mosul.

Place Name	Coordinates		Reference	Remarks
Sinope (text: Sinūb)	64;0, (35;9,	47;0 42;2)	74v:6,10	On the Black Sea coast of Turkey, west of Samsun.
[S]īrāf, called Shīlāb	89;0, (52;20,	29;0 27;43)	73r:3,88	Mod. Ṭāherī (*Bat.*, vol. 2, p. 43, no. 175; *LS*, p. 258) on the Gulf coast of Iran across from Baḥrayn.
Sirjan	90;20, (55;44,	29;30 29;28)	73r:3,93	Mod. Saīdābād (*Bat.*, vol. 2, p. 53; *LS*, p. 300) in southern Iran, SW of Kerman.
Sirrayn	77;15, (40;40,	20;0 19;37)	72v:1,20	In the Yemen.
Siṭīf	37;15, (5;24,	31;0 36;11)	73r:3,12	Mod. Setif, Algeria
Sīvās	71;30, (37;1,	40;10 39;44)	74r:5,30	In central Turkey, SSE of Samsun.
Siyāh Kūh Island	89;0,	43;30	74r:5,46	According to *Abū'l-F.* (transl., vol. 2, pt. 1, p. 326) this is an island in the Caspian.
Slavs, see Ṣaqāliba				
Sofāla Zanj	60;0, (34;52,	2;30 −19;49	72v:0,5	Other sources give the latitude as 2° below the equator. In Mozambique on the east coast of Africa.
Ṣohār	84;0, (56;45,	19;20 24;23)	72v:1,26	In Oman
Somnath (text: Ṣanamsumnāt)	107;10, (70;31,	22;15 20;50)	73r:2,42	In western India, on the southern coast of Gujarat.
Sudan, see Bulayda				
Ṣūdāq beside the Niṭūsh (Black) Sea	66;0, (34;57,	51;0 44;52)	74v:8,3	Mod. Sudak, Crimea.
Sūfāra	114;15,	19;35	72v:1,33	In India ?
Suhraward	83;20,	36;0	73v:4,93	The town, which no longer exists, was in Iran, between Zanjān and Qazvīn (*LS*, p. 223). Other sources give the same coordinates for Shahrūd, in Khurāsān, probably erroneously.
Sūkchū	124;0,	40;0	74v:5,77	In western China (*Ḥud.*, p. 232).
Sulāb (?) Island	88;30,	25;0	72v:2,35	In the Persian Gulf.
Sulghāt, being Qrim (the Crimea)	67;10, (35;6,	50;10 45;3)	74v:7,7	Sometime Eskikrim, Starykrim, mod. Simferopol (*Abū'l-F.*, transl., vol. 2, pp. 1, 320).

Place Name	Coordinates		Reference	Remarks
Sulṭāniya	84;0, (48;50,	36;30 36;24)	74r:4,102	In NW Iran, west of Qazvīn.
Sumayṣāt	72;15, (38;32,	37;30 37;30)	73v:4,48	Mod. Samsat, in southeastern Turkey, on the Euphrates north of Urfa.
Suquṭra Island	84;30, (53;59,	[1]3;0 12;40)	73r:3,75	Text has lat. 33. Mod. Socotra, in the Indian Ocean off the NE tip of Africa.
Ṣūr, see Tyre				
Surt	57;15, (16;39,	31;15 31;10)	73r:3,25	Also Sirte, in Libya.
Sūs al-Aqṣā	15;30, (−8;35	24;0 30;31)	72v:2,1	Mod. Taroudannt, Morocco.
Sūsa	44;15, (10;38,	32;30 35;50)	73r:3,18	Mod. Sousse, Tunisia.
Suwaydiya	71;0, (35;55,	36;0 36;7)	73v:4,40	Mod. Samandag, ancient Seleucia, the seaport of Antioch (*Abū'l-F.*, transl. vol. 2, pt. 2, p. 12).
Ṭabas-i Gīlakī	92;0, (56;54,	33;0 33;37)	73r:3,95	Mod. Tabas (*LS*, pp. 359–363), in east-central Iran, a hundred miles NW of Bīrjand.
Ṭabas-i Sīnā	94;0, (60;14,	33;15 32;48)	73r:3,100	Mod. Tabas (*LS*, pp. 359–363), in east-central Iran fifty miles east of Bīrjand.
Tabrīz	82;0, (46;18,	38;0 38;5)	73v:4,85	Capital of Iranian Azarbayjan.
Tadla (text: Tālād)	22;0, (−6;18,	30;15 32;34)	73r:3,6	Mod. Kasba Tadla, Morocco.
Tāhart, Lower	36;0,	29;0	73r:3,11	In the Maghrib.
Tahārt, Upper	35;30,	31;45	73r:3,10	In the Maghrib.
Ta'if	77;30, (40;21,	21;20 21;15)	72v:2,25	In Arabia.
Ta'izz (text: Ḥisn Ta'izz)	75;30, (44;2,	13;40 13;35)	72v:1,16	In northern Yemen.
Takīnābād	101;15,	33;0	73v:3,109	A place in Zābulistan, southern Afghanistan; a misreading of Bakrābād (*LS*, p. 347).
Tālād, see Tadla				
Ṭālaqān-i Khurāsān	98;15, (64;13,	37;30 35;43)	74r:4,140	Mod. Qaysar, in NW Afghanistan east of Bālā Murghāb (*LS*, p. 424).
Ṭālaqān-i Tukhāristān	102;50, (69;29,	37;25 36;46)	74r:4,149	In Afghanistan, east of Balkh. *LS*, p. 428, says the better form is Ṭāyiqān.

Place Name	Coordinates		Reference	Remarks
Tāna	102;0,	19;20	72v:1,29	On the coast of India.
Tangier (text: Ṭanja)	18;15, (−5;50,	35;0 35;48)	73v:4,1	
Ṭarāz, called Nīkī(?)	99;50, (71;25,	44;31 42;50)	74v:6,21	Mod. Dzhambul (*LS*, p. 486), in the Kazakhstan SSR, NE of Tashkent.
Tarragona (text: Ṭarkūna)	38;0, (1;15,	43;0 41;7)	74r:5,17	On the Mediterranean coast SSW of Barcelona.
Tarsus	68;40, (34;52,	36;50 36;52)	73v:4,28	In south-central Turkey, west of Adana.
Ṭawāwīs	97;40, (64;29,	39;30 39;52)	74v:5,53	Mod. Galaassiya (*LS*, p. 259), in the Uzbek SSR, a northern suburb of Bukhārā.
Taymā	67;15, (38;30,	25;40 27;37)	72v:2,13	In NW Saudi Arabia.
Thule (text: Būdan), Islands of	——,	63;0	74v:8,5	This is a conjecture. For longitude 20° and the same latitude *Tus.* and *Ulg.* give Tawā (=Thule) presumably obtained from the *Handy Tables.*
Tiberias (text: Ṭabariya)	68;15 (35;32,	32;0 32;48)	73r:3,44	
Tibet	110;0,	40;0	74v:5,75	Kāshī, together with the other medieval sources, gives a latitude which is much too far north.
Tiflīs	83;0, (44;48,	43;0 41;43)	74r:5,40	Mod. Tbilisi, capital of the Gruziya (Georgia) SSR.
Tikrīt	78;25, (43;42,	34;0 34;36)	73v:4,71	In central Iraq, north of Baghdad on the Tigris.
Tilimsān	24;0, (−1;21,	33;40 34;53)	73r:3,7	Mod. Tlemcen, Algeria.
[Tinn]īs	64;30, (32;15,	30;40 31;15)	73r:3,34	An island on Lake Manzala, Egypt.
Ṭirnū	67;30,	50;15	74v:7,8	*Abū'l-F.* (transl., vol. 2, pt. 1, p. 318) is the only other source who gives this place. He says it is in the land of the Vlakhs (Wallachia). The longitude he gives is 47;30, of which Kāshī's is probably a corruption.
Tī[z], capital of Makrān	103;0, (60;14,	24;45 25;16)	72v:2,39	Just west of Chāhbahār in SE Iran (*Maq.*, p. 106)
Toledo (text: Ṭulayṭa)	20;40, (−4;2,	35;30 39;52)	73v:4,8	

Place Name	Coordinates		Reference	Remarks
Tortosa (text: Ṭurṭūsha)	32;30, (0;31,	40;15 40;49)	74r:5,10	On the Mediterranean coast of Spain, SW of Barcelona.
Ṭrābzūn, see Trebizond				
Trebizond (text: Ṭrābzūn, see also Furḍat al-Rūm).	73;0, (39;43,	43;0 41;0)	74r:5,31	On the eastern Black Sea coast of Turkey.
Tripoli of North Africa (text: Aṭrāblus Maghrib)	45;15, (13;12,	32;30 32;53)	73r:3,19	Capital of Libya.
Tripoli of Syria (text: Aṭrāblus Shām)	69;40, (35;50,	35;15 34;27)	73v:4,34	In north Lebanon on the Mediterranean coast.
Ṭulayṭa, see Toledo				
Ṭulmaytha	54;0, (20;55	33;10 32;42)	73r:3,24	In Libya, ancient Ptolemais.
Tūn	92;30, (58;9,	34;30 34;0)	74r:4,125	Mod. Ferdows (*LS*, p. 353), in eastern Iran, NNW of Bīrjand.
Tunis	42;30, (10;13,	33;31 36;50)	73r:3,17	Capital of Tunisia.
Tunkat	101;0,	43;25	74v:5,67	Capital of the Īlāq region in the valley of the modern Angren River in the Uzbekistan SSR. The exact site of Tunkat is unknown (*LS*, p. 483).
Ṭurra(?)	49;20,	19;0	72v:1,4	In the Maghrib.
Turshīz	92;0, (57;30,	35;0 35;0)	74r:4,122	In east-central Iran (*LS*, p. 354) north of Tabas.
Ṭurṭūsha, see Tortosa				
Ṭūs	92;30, (59;31,	37;0 36;30)	74r:4,124	In NE Iran, just north of Mashhad.
Tustar	84;30, (48;51,	31;30 32;3)	73r:3,72	Mod. Shustar, in SW Iran, cf. *LS*, p. 234.
[T]uṭīla	30;30, (−1;37,	43;15 42;4)	74r:5,8	Mod. Tudela in NE Spain.
Tūzir	47;30,	29;10	73r:3,20	In the Maghrib desert.
Tyre (text: Ṣūr)	68;30, (35;12,	32;40 33;16)	73r:3,47	
Ubulla	84;0, (47;49,	30;20 30;33)	73r:3,70	Mod. Maqil, in southern Iraq north of Baṣra.
ʿUkbarā	79;0, (44;16,	33;0 33;49)	73r:3,56	Mod. Dujayl, in central Iraq north of Baghdad.

Place Name	Coordinates		Reference	Remarks
Ūrmiya	79;45, (45;2,	37;0 37;32)	73v:4,75	Mod. Rezā'iyeh, in western Iran near the Turkish border.
Usrūshana	100;0, (68;59,	40;0 39;58)	74v:5,61	Mod. Ura-Tyube (*LS*, p. 474), Tadzhikistan SSR, south of Tashkent.
Uzkand	102;50, (73;14,	44;0 40;48)	74v:6,24	Mod. Uzgen (*LS*, p. 476), in Kirgizia, USSR, east of Andizhan.
Valladolid (text: Madīna Walīd	21;12, (−4;45,	43;3 41;39)	74r:5,4	In north-central Spain, NNW of Madrid.
Venice (text: Bunduqiya)	42;0, (12;20,	44;0 45;26)	74v:6,5	
Walwālij	102;20, (68;51,	36;0 36;47)	74r:4,147	Near mod. Kunduz in northern Afghanistan.
Wāsiṭ	81;30, (46;20,	32;20 32;12)	73r:3,67	In SE Iraq NNW of Baṣra.
[Y]ābisa	30;15,	38;30	73v:4,14	A city on Ibiza Island, which see.
Yamāma	81;15, (47;24,	21;30 24;10)	72v:2,29	In Arabia.
Yamkot, see Jamkūt				
Yanbu'	74;0, (38;4,	26;0 24;7)	72v:2,19	Mod. Yanbo, Arabian seaport.
Yazd	89;0, (54;22,	32;15 31;55)	73r:3,89	In central Iran.
Zabīd of Yemen	74;20, (43;18,	14;10 14;10)	72v:1,11	On the Red Sea coastal plain south of Hodeida.
Zā[baj] Island (text: Zāyaḥ)	95;0	15;0	72v:1,28	Perhaps Sumatra (*Ḥud.*, p. 228).
Ẓafār of Yemen	77;35,	13;20	72v:1,24	In the Tihāma.
Zaghāwa	66;0,	1;10	72v:0,10	Other sources give it as in Zanj, S of the equator.
Zam	99;0, (65;10,	3[8];35 37;53)	73v:3,105	Mod. Kerki in the Turkmeniya SSR, on the Oxus (Amu Darya) east of Marv (Mary).
Zamakhshar	95;30, (60;43,	41;0 41;35)	74r:5,51	According to *LS*, p. 454, this site is four miles east of mod. Urgench, in the Uzbekistan SSR.
Zāmīn	99;40, (68;25,	40;30 39;56)	74v:5,59	Mod. Zaamin (*LS*, pp. 9, 475) in the Uzbek SSR, south of Tashkent.
Zanj, see Qanbala, Sofāla, Zaghāwa, Berbara.				
Zanjān	83;0, (48;30,	36;30 36;40)	73v:4,91	In NW Iran, SW of Rasht.
Zaragoza, see Saraqusṭa				

Place Name	Coordinates		Reference	Remarks
Zarand	92;15, (56;35,	30;15 30;50)	73r:3,96	In east-central Iran, NNW of Kermān.
Zaranj of Sijistān	97;40, (61;53,	32;30 31;6)	73r:3,102	In Afghanistan, on the Iranian border near Zābol.
Zawīla	49;0, (15;5,	30;15 26;11)	73r:3,21	In Libya.
Zayla'	71;0, (43;30,	8;0 11;21)	72v:0,12	Mod. Zeila, Abyssinian port.
Zaytūn	154;0, (120;7,	17;50 30;18)	72v:1,32	Mod. Hang Zhou, China.
Zūzan	93;30, (59;52,	35;19 34;22)	74r:4,128	In eastern Iran, west of Herāt (*LS*, p. 358).

ANALYSIS OF THE TABLE

Distribution of Localities by Regions

An arrangement of entries in the table according to somewhat arbitrary geographical regions comes out as follows:

Iran	89
Central Asia	77
Europe (other than the Iberian peninsula)	50
Turkey	44
Arabia	42
North Africa (excluding Egypt)	36
Geographical Syria	29
Iberian Peninsula	29
Iraq	28
India	22
Africa, East and Central	20
Egypt	18
China	15

Dependence Upon Other Sources

A statistical measure based upon the coincidence of city latitudes has been worked out to estimate the degree of borrowings between sources. According to this test, Kāshī is relatively independent of all early geographical collections, including that of Ptolemy (fl. A.D. 150), the father of mathematical geography. A possible exception is the anonymous author of a *Kitāb al-Aṭwāl*, a secondary source known only through *Abū'l-F.*, which may be early. Kāshī has obtained many coordinates from Bīrūnī, and even more from Naṣīr al-Dīn al-Ṭūsī. In turn, and not surprisingly, the geographical tables of Ulugh Beg use Kāshī as a primary source. The coordinates engraved on the backs of Islamic astrolabes and quadrants in many instances have come from Kāshī and the group of late Persian zījes related to his.

Latitudes

Since the geographical latitude of a locality equals the altitude of the celestial pole above the local horizon, this coordinate is easily determined by observing the meridian transit of a celestial body. Hence considerable precision can be expected of medieval latitude determinations. To some extent this is borne out by the table. The mean of the differences between Kāshī's latitudes and the modern values (where they are known) is only four minutes of arc. A more reliable measure of precision, in which negative errors do not cancel positive ones, is the mean of absolute values of the differences. This is 1;15°, which is not very impressive. In extenuation, be it said that the author could hardly be expected to verify personally the coordinates of places ranging from western Europe to China. At the same time, there is no evidence suggesting that Kāshī, like Bīrūnī, observed the latitudes of places where he resided or visited. In his introduction he boasts that, while modeling his work upon Naṣīr al-Dīn al-Ṭūsī's *Zij-i Īlkhānī* [*Tus.*] he has effected drastic improvements in it. Nevertheless he accepts Ṭūsī's traditional latitude (admittedly quite good) for his own home town of Kāshān, and the same goes for the observatory site at Samarqand.

Longitudes

Unlike latitudes, longitudes have no naturally defined starting point. Kāshī's zero meridian, in principle, runs through the Fortunate Isles (*Jazā'ir al-Khālidāt*, the Canaries). In order to determine a norm for converting between these values and longitudes measured from Greenwich, the mean of differences between Kāshī's longitudes and the corresponding modern values was calculated. It turned out very close to 33;30° (cf. [*K&R*]), hence this is to be added to a Greenwich value to make it compatible with a longitude from the text.

In theory the determination of a place's longitude is even simpler than that of its latitude. For, by virtue of the earth's rotation, the longitude difference between two localities equals the difference in their mean local times, converted at the rate of fifteen degrees per hour. But in practice, in medieval times the lack of a time signal simultaneously perceptible at each of two stations made the problem very difficult. Bīrūnī (in the [*Taḥdīd*]) applied with considerable success a rule giving longitude difference as a function of the known latitudes of the two localities and the great circle distance between them. But Kāshī's own geodetic rule for longitude difference (discussed in the [*Trig.*]) is quite inaccurate. Hence it is not surprising that his reported longitudes are far inferior to his latitudes.

Coordinate Errors by Regions

It is illuminating to plot Kāshī's localities, using both his coordinates and their modern counterparts (normed for the shift of meridian), on a single rectangular grid. This yields a pair of points for each city, which by their displacement from each other indicate the direction and magnitude of the error in each ancient determination. These vary greatly, even in the same neighborhood, but some general observations can be made.

In the middle of the plot, in Iran, Iraq, and Syria, errors tend to be minimal. This is the case also for much of Central Asia, except that north of the fortieth parallel localities fall far northwest of where they should be.

The Caspian and the Caucasus are west of their proper locations. On the other hand, places in Afghanistan and India are located much too far to the east. Places in the western part of the Arabian Peninsula are east of where they should be, whereas for Egypt the error is reversed. Thus if the Red Sea was plotted, it would turn out too wide.

Southern Turkey looks good, but in the north, cities are located far north of their actual positions.

Italy is displaced west of where it should be, and the error is compounded in Spain and western North Africa. Apparently, although the Muslims partially corrected Ptolemy's overestimate in the length of the Mediterranean, they did not go far enough.

In northern Spain latitudes are too high, and in North Africa they tend to be too low. This is surprising, considering the number of astronomers in both regions.

The Table of Climate Bounds

Kāshī, like most Muslim geographers, divided the greater part of the northern hemisphere into the seven "climates" of classical antiquity (see [*Hon.*]). These were defined as follows. The beginning of the first climate is at the parallel of latitude at which the maximum length of daylight is twelve and three-quarter hours. The "middle" of the first climate is the latitude at which maximum daylight lasts thirteen hours. The end of the first climate (and the beginning of the second) is the latitude enjoying thirteen and a quarter hours maximum daylight, and so on. The half-climates advance northward, each bounded by a quarter hour increase in maximum daylight. The end of the seventh climate is characterized by a maximum daylight of sixteen and a quarter hours. Kāshī's text first lists places south of the first climate, and he puts at the end localities north of the seventh climate. We call these two zones Climates 0 and 8 respectively.

To convert a maximum daylight, d, into a corresponding latitude, ϕ, calculate

$$\phi = \arctan\left[\sin 15\left(\frac{d}{2} - 6\right) / \tan \epsilon\right],$$

where ϵ is the inclination of the ecliptic, taken by Kāshī as 23;30°. This relation, or its equivalent, was familiar to any competent medieval astronomer.

In the lower left hand corner of the last page of the geographical table, f. 74v, is Kāshī's table of climates. It is transcribed below. In the column of climates *b.* and *m.* stand for "beginning" and "middle" respectively. In the column of latitudes most of the entries have a second number underneath in parentheses. The latter is the precise result. Where none appears, Kāshī's result is exact to seconds of arc. The mean error in Kāshī's determinations is about twenty-four seconds.

Table of Climate Bounds

Climate		*d* Max. Day (hrs.)	ϕ Latitude
1	b.	12;45	12;42,38° (12;42,13°)
	m.	13;0	16;42,22° (16;42,33°)
2	b.	13;15	20;31,1° (20;32,16°)
	m.	13;30	24;9,53°
3	b.	13;45	27;34,11° (27;34,31°)
	m.	14;0	30;45,35° (30;45,47°)
4	b.	14;15	33;43,5° (33;43,39°)
	m.	14;30	36;28,0° (36;28,27°)
5	b.	14;45	39;0,42°
	m.	15;0	41;20,5° (41;21,5°)
6	b.	15;15	43;30,20° (43;30,21°)
	m.	15;30	45;29,0° (45;29,18°)
7	b.	15;45	47;18,24° (47;18,43°)
	m.	16;0	48;59,20°
	b.	16;15	50;30,51° (50;31,55°)

BIBLIOGRAPHY

Abū'l-F.: M. Reinaud and MacGuckin de Slane, *Géographie d'Aboulféda*, texte arabe, Paris, 1840; . . . traduite de l'arabe en français, vols. 1 and 2. pt. 1, M. Reinaud, 1848; vol.2, pt. 2, Guyard, 1883.
AS: Aya Sofya (Istanbul) MS 2692, a copy of the *Zīj-i Khāqānī*.
Bat.: C. A. Nallino (editor and translator), *Al-Battānī sive Albatenii Opus Astronomicum*, 3 vols., Milan, 1899–1907.
Bir.: Bīrūnī, *Al-Qānūn'l-Mas'ūdī*, 3 vols., Hyderabad-Dn., 1954–56.
EI: *The Enclyclopaedia of Islam*, 4 vols., Leyden and London, 1913–34.
Guide Bleu: Moyen Orient, Paris: Hachette, 1953.
Hon.: Ernst Honigmann, *Die sieben Klimata . . .*, Heidelberg, 1929.
HT: N. Halma, editor, *Commentaire de Théon d'Alexandrie sur les tables manuelles astronomiques de Ptolemée*, Paris, 1822–25.
Ḥud.: Vladimir Minorsky (transl.) *Ḥudūd al-'Ālam, the Regions of the World, a Persian Geography . . .*, London, 1937.
India: Bīrūnī, *Kitāb fī taḥqīq ma li'l-Hind . . .*, ed. by C. E. Sachau, London, 1888; ed. Hyderabad-Dn., 1958; transl. by C. E. Sachau as *Alberuni's India*, 2 vols., London, 1910.
IO.: India Office (later Commonwealth Relations Office) Library, MS Ethé 2232, a copy of the *Zīj-i Khāqānī*.
Khu.: Hans von Mžik, ed., *Das Kitāb Ṣūrat al-Ard des Abū Ğa'far Muḥammad ibn Mūsā al-Huwārizmī*, Leipzig: Harrassowitz, 1926.
K&R: E. S. Kennedy and Mary H. Regier, "Prime Meridians in Medieval Islamic Astronomy," *Vistas in Astronomy*, 28 (1985), London, pp. 29–32.
LS: Guy LeStrange, *Lands of the Eastern Caliphate*, Cambridge, 1905.
Maq.: S. Maqbul Ahmad, *India and the Neighboring Territories in the Kitāb nuzhat al-mushtāq fī'khtirāq al-āfāq of al-Sharīf al-Idrīsī*, Leiden: Brill, 1960.
Taḥdīd: E. S. Kennedy, *A Commentary upon Bīrūnī's Kitāb Taḥdīd al-Amākin . . .*, American University of Beirut, 1973.
Trig.: E. S. Kennedy, "Spherical Astronomy in Kāshī's Khāqānī Zīj," *Zeitschrift für Geschichte der Arabisch-Islamischen Wissenschaften*, Bd. 2, 1985, pp. 1–46.
Tus.: Naṣīr al-Dīn al-Ṭūsī, *Zīj-i Ilkhānī*, Bodleian (Oxford) MS Hunt 143.
Ulg.: Ulugh Beg, *Zīj-i Sulṭānī*, Bodleian (Oxford) MS Marsh 396.

PLATE 1. India Office MS 430 (Ethé 2232), f. 72v. The beginning of the geographical table, Climates 0, 1, and 2.

فيه جدول اطوال البلدان من جزائر الخالدات وعروضها من خط الاستواء

PLATE 2. India Office MS 430, f. 73r. Climates 2 (cont.) and 3.

بقيه جدول طول البلدان عن جزائر الخالدات وعروضها عن خط الاستواء

اقليم چهارم

PLATE 3. India Office MS 430, f. 73v. Climates 3 (cont.) and 4.

بقية جدول اطوال البلدان عن جزائر الخالدات وعروضها عن خط الاستواء

PLATE 4. India Office MS 430, f. 74r. Climates 4 (cont.) and 5.

PLATE 5. India Office MS 430, f. 74v. Climates 5 (cont.), 6, 7, and 8, and the table of climate bounds.

CONTENTS OF VOLUME 77

Attention: Subscribers to the Transactions of the American Philosophical Society.

Please note the following corrections to Volume 77, Pt. 4 Alice Stroup. *Royal Funding of the Parisian Académie Royale des Sciences during the 1690s.*

ERRATA

Page 14, n. 5. and page 166, "Guy Picollet" should read "Guy Picolet"
Page 15, line 6 from the bottom of the text, "25 percent" should read "24 percent"
Page 17 (figure 2.2), line 1 from top, "6, 1, 6, 1, 7, 2, 1, 18, 24, 75" should read "6, 1, 6, 1, 8, 3, 1, 19, 25, 76"
Page 164, under "La Hire, Gabriel-Philippe de," omit "73-74," and "84"
Page 164, under "La Hire, Philippe de," add to "pension of," 73-74, 84," and under "professor royal," add "73-74"

www.ingramcontent.com/pod-product-compliance
Lightning Source LLC
Chambersburg PA
CBHW050350110426
42812CB00008B/2425